CHEMINS

I0074882

8 Uz
3293

Imp. ERHARD Frères, Rue Denfert-Rochereau, 35 ———— Mont-Blanc

The P.-L.-M. Railway

The Paris-Lyon-Mediterranean railway is one of the most important of the world It runs over more than a quarter of the territory of France and reaches a development of nearly 10.000 kilometers. The receipts exceed 440 millions of francs a year. Crossing through regions of exceeding fertility whose aspect alone reveals their resources and their prosperity, the railroad of the P.-L.-M. joins with Paris and with one another such commercial and industrial centres as Lyons, the se-cond town of France, Saint-Etienne, the Creusot, etc., all of first rank, and finally Marseilles which innumerable lines of steamboats put into communication with all the big ports of the Old World and the New.

From the picturesque point of view the P.-L.-M. railway is one of the richest of Europe. It passes through towns of immense archaeological inte-rest : after Fontainebleau, Sens, Dijon, Vienne, come the quaint old cities of Orange, Avignon, Arles, Nimes, etc., whose numerous monuments still recall to us Rome the mighty civilizer.

It is the P.-L.-M. railway the traveller must take when he wishes to start off on any of the best known excursions : Geneva, the key of the Alps; Chamonix, at the foot of Mont-Blanc; Grenoble and the Grande Chartreuse, the Vercors and the imposing glacial mass of the Meije and the Pelvoux.

It runs as well into the two queens of thermal stations, Vichy and Aix-les-Bains, without counting a quantity of excellent watering places : it stops at : Pougues, Royat, Châtelguyon, in the midlands; Vals in the Cevennes; Thonon, Evian, on Lake Leman ; Uriage and Allevard, among the beautiful Alps of the Dauphiné.

Again, it is the P.-L.-M. railway that the tourist takes to reach Switzerland, travelling with the greatest amount of comfort to Neuchatel, Lausanne and all the well known places on lake Geneva (Vevey, Montreux, Clarens) or again to Berne, to Thoune, to Interlaken, « the pearl of the Oberland », to Martigny (Great Saint-Bernard), to Zermatt, at the foot of the terrifying Cervin and quite close to the immense glaciers of Mount Rosa.

And when winter and the cold days come it is still by the P.-L.-M. line that the tourist will be carried into the enchanted regions of the Mediterranean, from Hyères to Mentone, stopping at Cannes, at Nice, at Monte-Carlo ; or, further still, into Italy, towards Milan, Turin, Lake Major, and Lake of Como, Venice, Genoa, Pisa, Florence with its celebrated museum, immortal Rome, Naples, Pompéii, or into the more distant regions of Algeria, of Sicily, of Egypt, of the East.

FONTAINEBLEAU

By the splendou
of its castle as much
as by the incomparable beauty
of the forest around it, is Fontainebleau regarded as one of
the most delightful excursions in the
vicinity of Paris.

An hour by rail and the traveller
finds himself in an old aristocratic
town. After a glance at the monument
raised in memory of President Carnot
you must start by visiting the celebrated castle of Fontainebleau full of
historic memories. Here were born
Henry III and Louis XIII; here the
great Condé died. But above all are we
reminded of Napoléon I, for it was
here he loved to come and rest in the intervals
of his campaigns. His rooms may still be seen, his
throne and the table on which he signed his abdication. The palace contains a quantity of interesting halls and chambers decorated with the
magnificent work of Primatice, Watteau, Boucher, and hung with Beau-
vais and Gobelin tapestries and priceless Lyons hangings.

Fontainebleau is surrounded by the finest forest of France. Here may
be seen remarkable trees : oaks, hundred of years old, preserved and

cared for like real monuments. The forest, full of diversity, abounds in famous spots : The *Roche-qui-pleure*, the *Caverne des Brigands*, the *Gorge aux Loups*, the *Mare aux Fées*, the *Dormoir;* bare and sunburnt rocks, deep gorges, smiling valleys. Just as above the sombre tree tops it is not unfrequent to see birds of prey so, too, in the coppices the tourist often meets with the stag and timid hind; and the spectacle is one of an unforgettable charm.

All this magnificent forest breathes a splendour of nature in which our greatest painters have found their inspiration.

It was amid the plains of Barbizon that J. F. Millet painted his *Angelus*. It was in the *Gorges de Franchard*, the *Gorges d'Apremont*, among the trees of the *Bas-Bréau*, so serene and lofty, that Théodore Rousseau. Corot, Daubigny and many others conceived their immortal masterpieces.

SENS attracts the travel-
ler's attention by
its remarkable cathedral and by several old houses, of
quite an individual Renaissance type. The cathedral of
Sens, *Saint-Etienne*, is a Gothic church of the XIIth century. It has been repaired several times and has this
curious feature that it has neither transept nor apsidal
chapel. The façade with its two spireless towers is
austere in spite of its multiple sculptures. It is said it
was here that the architect, William of Sens, invented the ogive ; it was he
also who built the celebrated cathedral of Canterbury in England. If the
exterior of *Saint-Etienne* seems somewhat heavy, to balance this defect
the interior is admirable with its vast nave, whose religious spirit instantly
seizes the visitor. Among the things to examine here, are the fine mausoleum of the Dauphin, the father of Louis XVI, with its four marble
statues by Coustou and the mausoleum of cardinal Duprat.

The treasury of the cathedral of Sens is the wealthiest church treasury
of France ; it is rich in tapestries, chasubles, gold and silver work of an
exquisite art. This treasury contains real marvels amongst which may
be pointed out to the traveller an ancient reliquary in gold incrusted with
precious stones and a superb ivory Christ by Girardin.

Sens possesses a museum containing a considerable Gallo-Roman
lapidary collection which recalls to us the fact that this pretty town rising
upon a picturesque and varied site, was a Roman favourite. Here they
built several villas of which many ruins still exist.

DIJON

The ancient town of the Dukes of Burgundy; today a pretty little city, clean and coquettish which does a great trade in wine. Dijon is proud of its remarkable monuments: first, its Cathedral, *Saint-Bénigne*, a Gothic church with two fine towers and a crypt of the XIth century;

the *Hôtel de Ville* where the tourist may examine a singularly valuable museum and will gladly stop to admire the Holbeins, the Rubens, the Ruysdaëls, the Teniers, etc. Dijon possesses as well many interesting old houses dating from the Renaissance: the *Hôtel Vogüé*, the *Maison Milsand*, the *Maison Richard*, with its courtyard and wooden gallery, are all charming **evocations** of the past.

7

LYON

Lyons, situated at the confluence of the Rhône and the Saône, may be regarded as the second town of France; and that, not only on account of its population and its extent, but also on account of its industry and commerce.

To understand to the full its beauty and its importance, the traveller should behold it from the heights of Notre-Dame de Fourvières. Here the meandering course of the Saône, there the rapid rush of the Rhône; here an ocean of houses, above whose mass emerge the towers of many a church; there, beyond the town, the Mont-d'Or, the mountains of the Chartreuse, the massive pile of the Pelvoux and finally at a distance of 93 miles Mont Blanc mantled in its eternal snows.

Lyons takes the first place in the world for its silk factories; it turns out more than 400 millions worth a year; everywhere are its incomparable silks admired and nowhere can they be imitated. — The *Place Bellecour*, the *Place des Terreaux* in front of the Hôtel de Ville, the *Place Carnot*, the big park of the *Tête d'Or* combine to make Lyons one of the pleasantest of towns. Monuments of great interest also abound. The church of Ainay which dates from the XIth century; St-Jean, the Cathedral, of the XIIth century, built at the foot of the hill of Fourvières, and at the top of this hill in the quite modern church of *Notre Dame-de-Fourvières* whose interior is richly embellished with marble columns and mosaics. A fine Town-Hall, an important Museum, a frequented public Library and spacious theatres complete an imposing whole.

It is not astonishing that a group of artists and polished writers should have dreamed of making a kind of French « Bayreuth » of Orange and of convening hither all lovers of high Art. Here in truth, and only here, is to be found the suitable frame for the masterpieces the old civilisations have bequeathed us; here only is modern talent able to reveal itself in a marvellous unfolding of scenery and under conditions of the most perfect fidelity to tradition. It was in the *Roman theatre* of Orange that Sarah Bernhardt won one of her greatest triumphs and that Mounet-Sully in *Œdipus Coloneus* made us shudder at the terrible tragedy of Sophocles. Once more are we impressed by the eternal power of Rome in the presence here, at Orange, of the *Arch of Triumph* of Marius, with its three arcades and its Corinthian fluted columns of supreme elegance.

AVIGNON If you behold Avignon from the banks of the Rhône, it presents at a first glance the grand and imposing pile of the *Palace of the Popes*, which arrests the eye. Round the town, almost intact, are still the old *city-walls* which carry us back to the middle ages. Avignon, once the town of the Popes, has become the heart of the Provençal Revival. There are some interesting churches to be seen, *Notre-Dame-des-Doms, Saint-Symphorien et Saint-Pierre.* From the *Promenade du Rocher des Doms*, which rises abruptly above the Rhône, the traveller enjoys a view of the valley of the Rhône with the famous bridge of Bénézet, the bridge « where they dance round dances », as the old nursery song sings.

NIMES No town of France offers the traveller more numerous vestiges of the ancient civilisation of Rome, nor monuments in a better state of preservation. It is with surprise that we first behold the Arena of Nîmes, which held as many as 30 000 spectators and is in more perfect preservation than the *Colosseum of Rome* itself.

But one of the loveliest gems of a forever vanished art may be contemplated in the famous *Maison Carrée*, a Roman temple of an exquisite purity of style, a treasure that would suffice in itself to render forever illustrious the town that possessed it.

NIMES

Arles

It is Rome itself that still lives in this little sunlit town of Provence. Arles takes pride in its arena, one of the largest amphitheatres that the Romans built in Gaul, the boldness and grandeur of whose architecture strike the eye with awe. At Arles also may be admired the eloquent ruins of an *antique theatre*. In contrast with this Pagan art it is curious to note, adjoining the cathedral *Saint-Trophime*, a romanesque cloister with delicate little pillars, breathing a marvellous spirit of religion. But it would seem that the loveliness of Greece has stamped something of the eternal youth of its beauty upon the Arlesian women with their visages of so pure and charming a type, — those pretty Arlesian women, Alphonse Daudet has hymned immortally, women of a healthy and vigorous grace, who have kept to our days their quaint and delightful costume.

MARSEILLE

Marseilles « the Gate of the East » such is the name of one or the magnificent frescoes of Puvis de Chavannes, which decorates the staircase of the Palace of Longchamp. Marseilles is indeed the gate which opens into these enchanted lands, into these mysterious and sun washed countries of the East. Here already end the regions of cloud laden skies; the air is more vibrant, the light more intense. You need but to watch a departing ship, to follow it on its outward way with your eyes to feel that already the voyage towards lands of palm and oleander has begun.

The spirit of reverie seizes you as you stand on the edge of the blue washed Mediterranean. These same waves kiss Italian shores; Greece is near and you are reminded that the Phoceans founded Marseilles; for long was it known as « New Athens ». All the glory of a magnificent antiquity is evoked for us ; the most splendid facts in the history of peoples took place upon the shores of the unique Sea. Below Egypt with its great unmovable gods, its Pyramids, its tombs; further still, Constantinople with the Golden Horn; ancient Byzantium, of magic name. Again is Marseilles the open door to India, China and Japan. The formidable past of humanity greets us here as well as the impression of the colonising and commercial activity of the races of to-day. We feel that a new life begins here and continues on far from the harbour, and that in the immense docks of la *Joliette* the heart beats that will carry as far as the East, on a wave of generous blood, the renewed life of our civilisation.

Seen from the sea, Marseilles presents a unique spectacle, built amphitheatre-wise on the hills that encircle the bay, on one of which rises the *Chapel of Notre Dame de la Garde*. The neo-byzan- tine Ca-

thedral should be visited; its interior is very striking; also the *Palace of Longchamp*, of which mention has already been made which, along with the Opera of Paris, may be regarded as one of the finest monuments raised in France in the course of this century. The *Palace of Longchamp* is built in Renaissance style and its architectural lines are of a rare felicity and a perfect grace. Two great constructions rise on either side and are connected by a harmonious colonnade in hemicycle adorned with a triumphal arch which is the Château ...eau or aqueduct of Marseilles. In front a graceful cascade and a colossal group sculptured by Cavelier representing « la Durance » drawn by four bulls. The *Palace of Longchamp*, as a whole, presents a magnificent aspect. It possesses one of the most interesting Provincial Museums. We have mentioned the frescoes of Puvis de Chavannes which decorate the staircases; there are also some fine Jordaens, Raphaëls, Rubens and important specimens of the Dutch and Lombardy Schools.

Marseilles is above all a town full of animation and gaiety. Its climate is delicious and its population, of brisk and sprightly characters contributes not a little to render a stay there agreeable. The town is laid out in superb boulevards and avenues. Many of them are well known : the Alleys of Meilhan shadowed by their big trees, the rue Noailles and the famous *Cannebière*, the pride of every Marseillais. The Cannebière is a wide boulevard which runs down to the Old Harbour. It is bordered by luxurious dwellings and brilliant cafés. They make this pretty corner of Marseilles in the warm lights of sunset, with the forests of masts of the Old Harbour beyond, one of the pleasantest of spots, which we leave regretfully.

On the right hand of the Cannebière is the vast and handsome building, the *Bourse*, with a loggia ornamented with a bas relief which represents Marseilles receiving the produce of the entire world. The Chamber of Commerce of Marseilles meets at the *Bourse* and is one of the most important and wealthiest of the world.

Behind the *Bourse* spreads the *Old Town*, as far as the *Town-Hall* which stands in sight of the *Old Harbour*. The quaint and curious quarters of the fishers-folk are interesting to visit.

At the end of the Old Harbour, the *Fort Saint-Jean ;* in front the *Château du Pharo* and the vieille *Abbaye of Saint-Victor*.

A few steps beyond the Cannebière rises the monumental lift in connection with the hill and the new chapel of *Notre Dame de la Garde ;* this sanctuary, so popular with Provençal sailors, overlooks the entire town. No thing could be finer than the view from Notre Dame de la Garde ; the panorama of the whole town ; the harbour and islands ; seven miles from the coast, beyond the Château d'If, the tower of Planiers, — one of the most powerful lighthouses of the Mediterranean.

We must also point out the fine Promenade of the *Prado*, which comprises a magnificent avenue of more than three kilomètres (2 and a half

miles). It is much frequented and at the end of it may be seen the *Château Borely* which contains a valuable museum of Marseillais, Egyptian and Phœnician antiquities.

The Promenade of the *Corniche* claims us ; the way is cut through rocks and offers us superb views over the Bay of Marseilles, above all of the *Château d'If* which Alexandre Dumas has immortalized in *Monte-Cristo*: it is upon this route, a little before reaching the handsome Promenade du Prado you will find the well-known restaurant, famous for its Marseillais dish, the Bouillabaisse (sung by Thackeray), in front of an admirable view of sea and rocks.

Marseilles is the first commercial port of France. About 17.000 ships of every land and tonnage carrying in their flanks nearly six millions of tons of every kind of merchandise enter it yearly. The port of Marseilles, of all French ports, receives the greatest quantity of cereals more than a million tons enter France through Marseilles.

A visit to the *Docks of la Joliette* easily gives us an idea of its prodigious importance. It is there that all the big steamers of the world drop anchor. Every year the huge liners convey more than 300.000 passengers, every year 1624 ships leave the docks of Marseilles for Algeria and Tunisia ; 130 for Greece ; 312 for Turkey in Europe and in Asia ; 275 for Egypt : 160 for India : 104 for China and Japan ; 65 for Australia ;

2

MARSEILLE

108 for South America; such figures constitute an incontestable eloquenc[e]

But Marseilles is something more than a large commercial town ; i[ts]
rank in France is that of an actively industrial city, by its soap, oil, an[d]
candle factories, its tanneries and sugar bakeries.

This will explain why Marseilles is the most active and animated [of]
centres. Life perpetually flows through it, and the immense labour of th[e]
city is ended in the evening by a joyous invasion of streets and prome[-]
nades. When in the midst of its active population we meet with the foreig[n]
and multicoloured garments of men and women from far off lands, th[e]
strange, original, unexpected look of Marseilles, where the movement [of]
a large and wealthy city is wedded to the smiling charm of a sunlit an[d]
mirthful town, gives us the impression of a place which has had th[e]
grace to preserve from the olden times of its heroic origin a little [of]
the charm of the East it opens to us and of which it is the presage.

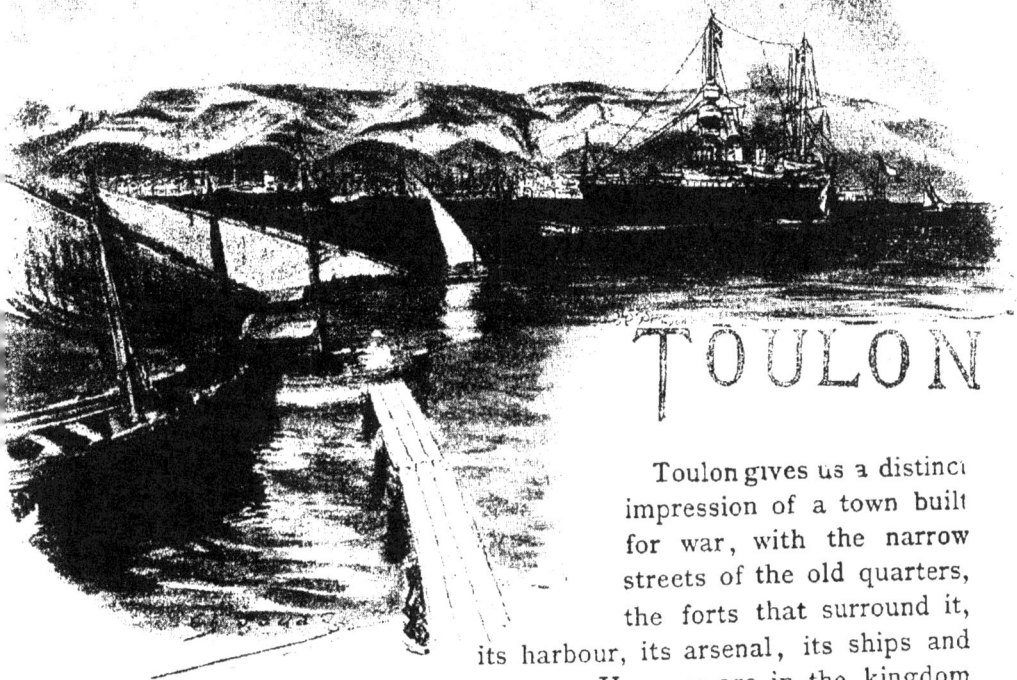

TOULON

Toulon gives us a distinct impression of a town built for war, with the narrow streets of the old quarters, the forts that surround it, its harbour, its arsenal, its ships and cannons. Here we are in the kingdom of Force, of an admirable, disciplined force. Nothing could be more striking than the spectacle of the squadron in the Mediterranean and a visit to a man-of-war leaves us profoundly conscious of the immense might hidden in the flanks of these prodigious engines. The Cathedral of Toulon, *Sainte-Marie-Majeure* is extremely interesting; the *Hôtel de Ville*, the *Port* and the magnificent *Place d'Armes* upon which stands the *Préfecture Maritime* combine to make Toulon one of the finest fortified places of France. In the vicinity, *Tamaris, Hyères* and *La Seyne* are delightful excursions.

CANNES

Cannes has been compared to a snow-white seagull stretched lazily on the edge of the azure sea. No other spot on the Riviera gives such a vivid impression of a charming and aristocratic town where it is good to dwell awhile in luxury and repose. If you look at Cannes from the sea, it offers a sunlit front of hundreds of white villas surrounded with palm and orange branches, all gay

and smiling with behind the protection of an amphitheatre of hills and further inward still the outer chain of the Alps a white blaze of snow. From the splendid walk of the Croisette in the early sunrays it is an unforgettable fairy-scene of opal and sapphire waters throbbing, softly lifted by the rhythm of some divine breathing. Surely it can only be here that the poet was able to see :

.....Vénus Astarté, fille de l'onde amère.
Secouant, vierge encore, les larmes de sa mère
Et fécondant le Monde en tordant ses cheveux.

Nice

Gaiety, a prodigious
animation, dazzling sun-
shine, the scent of
flowers, charming wo-
men, an exquisite cli-
mate, palms, sapphire
sea, and all in the midst
of comfort and the best
of living, this is Nice,
Nizza la Bella, whose
fame extends over all
the world. From every
capital at the first breath of winter, the
world of elegance flies to Nice. Nice is the smile of
France. The traveller leaves London or Paris dim with fogs, mantled
with snow or inundated with rain, and steps out of his train in the heart of roses,
breathes an air smelling of violets and mimosa, along the fairy Promenade des Anglais, in the
midst of a brilliant and fashionable crowd.

Nice abounds in varied pleasures : famous actors and actresses and many of our most
popular stars are, each winter, applauded in its theatres. Concerts take place continually and
masked balls, « Veglioni », of an irresistible animation. At Carnival time the famous *Battle of
flowers* is fought along the Promenade des Anglais; to and fro, in waves of sunlit dust.
All carriages garlanded from the wheels to the harness; and from vehicle to vehicle the best
society flings armfuls of flowers in the midst of the laughter and delight of a population in
holiday mood.

Sports prove equally attractive to the fashionable world of Nice : in January the Races
begin the season and the Regattas, in April, end the long series of festivities of this
enchanted town.

Besides the *Promenade des Anglais*, Nice possesses a *Public Garden* planted with
palms and oleanders where the *Monument of the Centenary* of the union of Nice with
France should be seen. We must also mention the *Place Massena* with its arcaded houses
and a quaint flower-market where nearly all the tourists come every morning to pur-
chase magnificent bouquets which they post off to their friends still breathing the mists of
the North.

The maritime quarter, with its pretty harbour of Limpia, is little known to strangers; it
possesses now no building to recall its old greatness. Yet this little fisher-village was the real
office of history, which alone bore the name down the Bourbon epoch.

We may point out also some delightful walks in the neighbourhood. In following the *Cours
du Paillon*, which has been covered again and upon which a handsome public garden has been

laid out, you come to the Monastery of Saint Barthélemy; further on is the castle of Saint André built on a remarkable site. On the road to Villefranche, walking through the forested path of Montboron, the traveller will enjoy a superb view above the roadstead of Villefranche, over Beaulieu and the lighthouse of Saint-Jean: the splendid highway of the Corniche, which runs from Nice to Mentone is one of the most enjoyable and picturesque of excursions. This road makes the turn of Mont Gros, where the Observatory of Nice stands, and beyond is seen the Alpine pinnacle bare and snow-hooded.

Such a place as Nice needed a first class establishment which would guarantee travellers comfort and superlative refinement in a matchless setting : such is the *Riviera Palace* at Cimiez, a neighbouring hill, where several handsome villas have been built. The hotel garden is planted with palm, orange and lemon trees; its beds are sheets of roses and violets; a thousand games are set up and every afternoon, at tea time, an orchestra plays. From the *Riviera Palace* Hotel the traveller enjoys the loveliest view above the marvellous panorama of the *Baie des Anges*, of the entire town and its vicinity ; it is a delightful residence.

MONTE CARLO

No town on earth is more marvellously situated; it enjoys together the mildest climate, the heavenliest of skies and the most captivating of landscapes. Lean but over the balustrade above the gardens of the *Casino* and you will have under your eyes a glorious panorama unfolded along the azure coast from the hills of the Estérel to the lovely coast-line of Ventimiglia and Bordighera. And so the fame of Monte-Carlo increases year by year, and tourists crowd hither from all parts of the world.

This town of fairy-land with its charming little streets, its luxurious houses, its wonderful gardens, is, in winter, the meeting place of all that is celebrated and elegant in social life.

Quite close to Monte-Carlo, the capital of the Principality of *Monaco* rises upon an abrupt rock. The Renaissance Palace, with its crenelated towers, possesses gardens of an incomparable beauty.

3

MENTON

A simpler and quieter place than Nice, Mentone is one of the chief winter residences of the Mediterranean.

Its climate is the warmest of the Riviera and the land abounds in orange, lemon, fig and olive trees. Its beautiful scenery and equable temperature make Mentone a haven of delight and of rest. Delicate persons will find here the yearned for spot where, far from the fatigues and cares of the world, health may be recovered. Above the harbour, on the hillside, rises the old town ; here you should see the church of *Saint-Michael* on a square whence the view spreads over a vast and magnificent panorama.

VICHY

Vichy was a familiar spot to the Romans; Madame de Sévigné sang the praises of its waters to Louis XIV; Napoléon III established its renown. To-day Vichy is the principal watering place of Europe; its springs the *Célestins*, the *Hôpital*, the *Grande Grille* are known all the world over. As well as this, the little town of Vichy is a gay and delightful place of residence. Its thermal establishment is a model of comfort and elegance and the Park, which separates it from the Casino, is one of the most frequented of meeting places. After the morning greetings of the « Buvette » come the afternoon promenades in the Park, where the visitors meet again, or the festive reunions of the large saloons of the Casino Life at Vichy is both quiet and lively. Numerous gaieties, admirable theatrical representations, charming excursions not only help the time under treatment to pass quickly but, better still, enable the bathers to forget that they are ill.

Aix-les-Bains

Aix is a cele-
brated thermal
station, known
of even in the time of
the Romans and which
has become an extraordinarily
fashionable resort. The town
is charming, finely situated at the foot
of *Mont Revard*, which may be ascended by the cog-wheel railway and
whence the tourist perceives Mont-Blanc like a gigantic rampart of ice.
Aix is frequented by a very select class of visitors for the curative qua-
lities of its warm springs of acknowledged efficacy in the treatment of
rheumatic cases of all kinds. The Thermœ with their sumptuously moun-
ted showerbaths and their famous massage saloons offer the visitors
every resource of modern medical science. The treatment is not however
so absorbing that the bathers are excluded from all the abounding amu-
sements got up for their entertainment, above all those of the two Casinos
which rival each other in attractions, and whither come every year the
most popular artists of the musical and dramatic world. Aix enjoys
an extremely mild climate, and the season begins in the Spring and ends
in the middle of Autumn.

Another charm of Aix lies in the excursions within its vicinity which
offers views of a remarkable beauty. We have mentioned *Mont Revard*;
but you have only to go outside the town to find yourself on the border
of the glorious *lake of Bourget* encircled by wonderfully picturesque
mountains. We recommend the excursion to the *Gorges de Sierroz*, the
ascension to the *Dent du Chat* and the *Pont de l'Abime*.

Situated upon the road to Italy, it is by thousands that Aix is visited.

ANNECY

In this ancient land of the counts of Geneva everything is picturesque, and in many respects Annecy continues to maintain its antique character. It is still dominated by its strong castle which has so imposing an air with its square towers and its machicolation. Some of the old streets of the town are still arcaded and run un- der vaulted passages. But the greatest attrac- tion of Annecy is its beau- tifullake spreading amidst a belt of

lofty and picturesque mountains. Numerous boats (that start from the beautiful promenade du Paquier) ply their way up and down and across it and enable the traveller to admire the villages that form themselves into charming and graceful groups of buildings and effects upon its borders : *Menthon*, *Talloires*, *Duingt* and its castle flanked by a round tower. We recommend also to the tourist the excursion to the quaint *Gorges du Fier*, and we may point out to him the ascension of the celebrated *Semnoz*, which has been compared with the Righi and from the heights of which may be witnessed fairy wonders of sunrise.

PONT SAINT-BRUNO. — LA GRANDE CHARTREUSE

GRENOBLE

LA GRANDE CHARTREUSE. — LA MEIJE

Not only is Grenoble a large and fine town. but it has a place among the most picturesque cities of France, because of its situation in the midst of lofty mountains which form a superb crown of snow around it. Day by day Grenoble increases its importance as a centre of excursions : at the town's gates is Uriage, a fashionable thermal station ; the « Moucherotte », whence may be seen the valley of the « Graisivaudan » washed by the Isère ; the « Grande Chartreuse », in the imposing group of the same name, the celebrated monastery, which may be visited in a day : you cannot enter it without emotion ; here far from the crowd are you enveloped in the silence of austere nature, in the meditation of cloisters whence are forever banished the clamours of life. We advise the ascent of the Grande Chartreuse by the Désert : a wonderful gorge cut between immense rocks and where we may point out the Bridge of Saint-Bruno whose bold arch spans the abyss.

Beyond Grenoble is the excursion to the Vercors, a wooded range of a striking aspect ; again, by the picturesque valley of the Romanche, the

La Meije

Bourg d'Oisans ; the glo-
rious glaciers of La Meije,
so mysterious and transpa-   rent, in the Pelvoux chain.
La Meije, that impressive height (3 987 mètres), may be compared with the
most celebrated mountains of Switzerland. Its ascension is difficult, but
the tourist can easily admire it from the little village of *La Grave* where he
will find himself in an almost matchless and marvellous glacial region. By
Grenoble you can also go to Aix-les-Bains, or to Savoy, of lovely scenery.

Geneva, the city of the Reformation; Geneva, where Calvin died and Rousseau was born, is one of the largest towns of Switzerland. It is admirably situated on the western side of Lake Leman, upon both banks of the Rhône which are joined by magnificent bridges. The widest is the *Bridge of Mount Blanc*. Between it and the Bridge des *Bergues* lies the *isle of Rousseau* where the statue of Jean Jacques stands. The traveller should visit the *Cathedral*, the *Town hall* and the *Rath museum* which contains some first-rate pictures.

Geneva is the Centre of unmatched excursions the world over. The towns along the shores of its incomparable lake are of universal reputation : Evian, Thonon, Clarens, Territet, Villeneuve, charming Lausanne, Vevey, Montreux and how many more that evoke visions of an enchanted land. Numerous steamboats run up and down the lake and permit the tourist to enjoy varied excursions that afford him views of the most delightful and changing of landscapes.

LAC LEMAN

The beauty
of Lake Leman has
inspired more than one
great poet. Voltaire and
Rousseau are among its sin-
gers. Its shores, with admirable views,
scattered round with villas and graceful villages, lend it an animated and smiling aspect. The Swiss side exposed to the South is of a
surprising fertility. The French side on the contrary is overlooked by
the high mountains of the Valais and Savoy. Boats plying constantly
to and fro furrow its blue waters and facilitate excursions which are a
perpetual enchantement.

Coppet where Necker lived; *Nyon* with its old castle ; *Rolle*, near
which is the Signal of *Bougy*, a magnificent panorama ; *Ouchy*, which
is the port of lovely *Lausanne*; *Vevey, Clarens*, with their charming villas ;
Territet, whence the funicular railway starts for *Glion*, which catches
up the railway of *Caux* and the *Rochers de Naye;* the *Gorge du Chau-
dron;* the *Avants ;* and, further still, the *Castle* of *Chillon* with its
silhouette of castellated fort that seems to dip into the waters; *Saint-
Gingolph, Meillerie* in a delightful spot; *Evian-les-Bains*, a very fashionable and frequented place; *Thonon, Yvoire*, and how many others !

Renowned all the world over, Lake Leman offers us the most splendid
and varied of landscapes in Europe, and attracts thousands of travellers

CHAMONIX

Chamonix is Mont Blanc. The giant of 4810 mètres is there, stupefying, grand and fascinating. Nobody can think of anything but it, nor talk of anything else. Few are those who have reached the summit, but all tourists dream of the glorious adventure. Nevertheless, caravans of courageous travellers continue year after year to file along the road to Mont Blanc by the Pass of the *Grands Mulets*, if it were only to walk knee-deep at the mountain base in the snows of the pinnacle. Happily it is not necessary to mount to the almost inaccessible summit of the most towering of the Alps to enjoy the sensation of an interesting excursion. There are those of the *Montenvers*, the *Chapeau*, the *Brévent* and the famous *Mer de Glace*.

ZERMATT

Zermatt, like Chamonix, offers us sublime pictures of the world of ice-fields. At Zermatt we are in the heart of the Alps; from the *Gœr-nergrat* which is near at hand, and which may be ascended by railway, the tourist discovers from an altitude of 3,200 metres an extraordinar belt of mountains, snow-mantled, which forms one of the most surprising spectacles the eye of man can embrace. Towards the south rise the imposing mass and the sharp peak of *Mount Cervin*. Its ascension implies some danger and the presence of an experienced guide is indispensable. The superiority of mountain views and sea-views is ever an open question; but what is incontestable is the fact that the mountains have a far greater attraction than the sea for the average tourist. Many who are content to regard tranquilly from shore the play of light along the iris-hued sheet of wavy water and contemplate its furies or the tragic horror of its tempest, are filled with impatience when they find themselves at the foot of a mountain. They must needs instantly arm themselves fearlessly with the irontipped stick and axe to risk an absorbing ascension At Zermatt you are the victim of a kind of mountain intoxication, you are « drawn towards the heights » where the light is more brilliant, the air is rarefied, is purer and lighter.

As soon as the traveller enters Berne he is struck by the medieval character of the town : arcaded houses, charming fountains, old corporation houses, terraces admirably disposed for the enjoyment of splendid panoramas Berne is the seat of the Federal Government and possesses an University. The traveller should see the Cathedral, the Federal Palaces, the famous Ditch of Bears (the bear has a place in the expressive armorial bearings of Berne), and above all should he go as far as the small rampart (Kleine Schanze) whence the high mountain range of the Alps may be seen above the town, which is mapped out, belted by the Aar. From Berne the railway takes you to *Thoune*, made famous by its magnificent lake, its castle and the superb peaks that encircle it.

Thoune

Interlaken et la Jungfrau

Here we find ourselves in the midst of Swiss wonders Interlaken, placed between the lovely lakes of Thoune and Brienz with the exceptional mildness of its climate, is the centre of marvellous excursions in the Bernese Oberland It ranks first among summer places. Several cog wheel railways climb the mountain flanks as high as Lauterbrunnen to cross to *Wengernalp* whence the *Jungfrau* is seen in all its magnificence.

The *Jungfrau* (the maiden) is one of the most celebrated mountains of Switzerland; its proportions are colossal and it rises to 4 167 metres It is an imposing mass, of striking poetry of line, and from its summit the most glorious sunrises may be seen. This ascension is attempted by hundreds of tourists. In a little while an electric line will run to the pinnacle by a subterraneous passage cut through the Mœnch and the *Jungfrau* : and travellers may contemplate from a glass pavilion, sheltered from the wind, an infinite number of peaks, of mountains and glaciers in a setting of incomparable scenery.

It needs but the vocable
of that town to recall to us
the purest memories of art,
the names of the most illustrious men, the most astonishing adventures of history. The whole of Florence
is a Museum; there is not a
street, a square, a monu-
ment which does not give the intensest impression of artistic and intellectual movement. The
ancient capital of Tuscany, the land of Dante,
the town of the Medici, Florence is enveloped
in an atmosphere of art and beauty which
makes her even to this day the educator of the
mind and the soul. To mention her monuments,
enumerate her palaces, her statues, her pictures, would be an unending labour; and the
little we could say would require long pages.
And how many masterpieces would be forgot-
ten! Among the foremost ones we stop at
the names of Michael Angelo, Raphaël, da
Vinci.

Venise

A shock of irresistible admiration and of profound wonder awaits the traveller upon his arrival at Venice. The Queen of the Adriatic occupies by her art the same unique and isolated place that she holds in history. Here canals take the place of streets, gondolas that of cabs and carriages. One of the most vivid impressions Venice leaves upon the tourist is the strange silence of the town. Here is not the place to make mention of the beauties of the marvellous city, not of *Saint Mark's church* and the *Piazza*, the *Palace of the Doges*, nor yet of its museums full of masterpieces, nor of the charm of its canals. Every poet has sung of Venice : every poet to come will sing of her.

MILAN

The capital of Lombardy, one of the most animated and wealthiest towns of Italy. It is a fine city, possessing eighty churches of which many are quite remarkable. What the tourist must see before anything else at Milan is the *Duomo*, its Cathedral. It is built of white marble, ornamented with 98 turrets and 2 000 marble statues; the interior, thanks to the dim light and the wonderful stained glass, produces a most striking effect. Near the Duomo, in quite modern contrast with it, is the *Gallery Victor-Emmanuel*. The traveller should visit the Romanesque church *Saint-Ambrosio*, *Santa Maria delle Grazie* where will be found the *Last Supper* of Leonardo da Vinci, *the Brera*, and finally the celebrated *Ambrosian Library*.

TVRIN

The traveller, in the glimpse of a first visit, may be surprised at finding a town built upon such a regular plan. The streets of Turin are laid out at right angles and this plan, we are told, dates from the period of the colonisation of Augustus. Turin is a bright and active town, crossed by splendid arteries. The lover of art can start his studies at Turin : the *Picture Gallery* contains an admirable Virgin of Lorenzo di Credi, a Memling, and interesting paintings of Van Dyck. Close to the Gallery is a fine statue of Emmanuel Philibert and on the « Piazza Carlina », the handsome *monument of Cavour;* the *Cathedral* is Renaissance. There are charming walks at Turin along the Pò, and we advise a visit to the *Campo Santo* and to the new monument lately raised to Victor Emmanuel. But above all, visit the *Superga,* a hill on the east of the town. Here there is an abbey and a remarkable church containing the tombs of the dukes of Savoy ; the tourist ascends by tramway and from the church a delightful view of the Alps may be had.

This illustrative pamphlet
has been issued by the

Paris-Lyons-Mediterranean Railway

Under the artistic care of
A. Romagnol

Paris. — MOTTEROZ, Lib.-Imp. réunies, 7. rue Saint-Benoît.

THE ALPS

IN CONNECTION

WITH THE P.-L.-M. SYSTEM

RÉSEAU P.L.M.

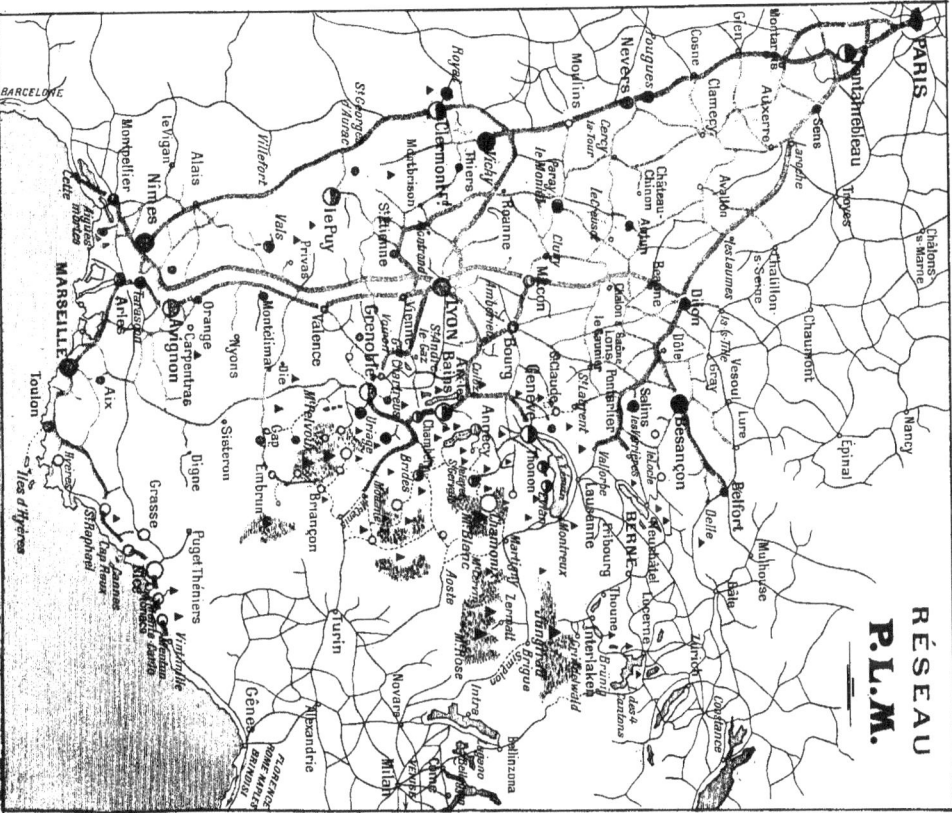

○ Centres d'excursions pittoresques (*montagnes, forêts, etc.*). — ● Stations thermales. — Attractions d'hiver. —

● Villes ou localités particulièrement intéressantes par leur importance, leurs monuments, etc.

▲ Principales montagnes. — ▭ Glaciers.

PARIS — Fontainebleau — Sens — Langres — Montargis — Gien — Auxerre — Clamecy — Nevers — Fougues — Cosne — Troyes — Chaumont — Châtillon s. Serge — Châtillon s. Seine — Vesoul — Lure — Belfort — Delle — Mulhouse — Bâle — Epinal — Nancy — Chalons s.Marne

Moulins — Cercy la-Tour — Château-Chinon — Avallon — Dijon — Dôle — la $\frac{1}{2}$ Lieue-le-Gray — Gevrey — Constance

BARCELONE — le Vigan — Montpellier — Cette — Aigues-mortes — Nîmes — Alais — St-Georges d'Aurac — Royat — Clermont-F. — Thiers — Vichy — Roanne — Pont-Gibaud — Mâcon — Cluny — Paray le-Monial — Beaune — Lons-le-Saunier — Salins — Pontarlier — Besançon — BERNE — Fribourg — Lucerne — Neuchâtel — les 4 cantons

MARSEILLE — Arles — Tarascon — Avignon — Orange — Carpentras — Nyons — Montélimar — Die — Privas — Vals — le Puy — St-Étienne — Montbrison — Vienne — St-Rambert — LYON — Bourg — Ambérieu — Genève — Aix-les-Bains — Annecy — Thonon — Montreux — Interlaken — Thoune — Brunig — Sernen — das 4

Toulon — Aix — Digne — Sisteron — Gap — Embrun — Briançon — Grenoble — Vizille — la Grave — Grasse — Puget-Théniers — Turin — Novare — Alexandrie — Milan — Gênes — Vintimille — Monaco — St-Raphael — Cannes — Hyères — Iles d'Hyères — Bellinzona — Lugano — VENISE — FLORENCE — ROME — NAPLES — BRINDISI — Mt-Blanc — Chamonix — Zermatt — Mt-Rose

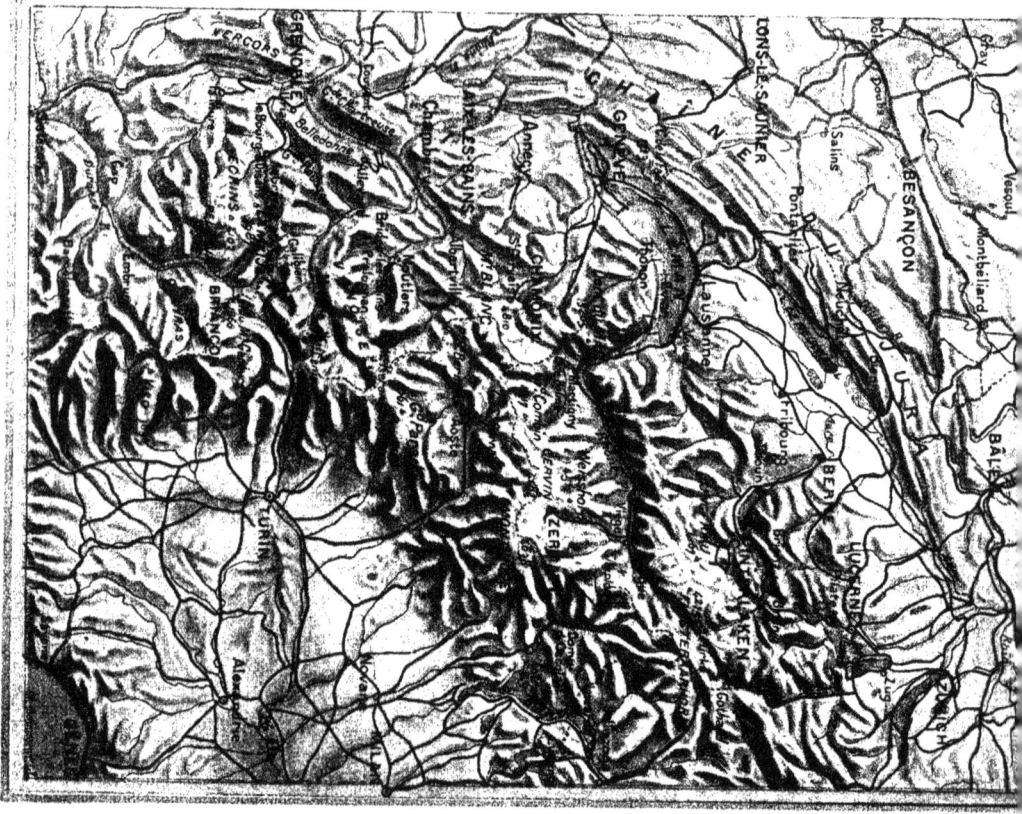

CONTENTS

MAPS

The Alps in connection with the P.-L.-M. system
Europe.
The P.-L.-M. system.

NICE

www.ingramcontent.com/pod-product-compliance
Lightning Source LLC
Chambersburg PA
CBHW071349200326
41520CB00013B/3165